Turn the Ship

A Justified Mutiny

Captain Tom McKeown USN

AquaZebra Publishing

© 2021 Captain Tom McKeown, USN

All rights exclusively reserved. No part of this book may be reproduced or translated into any language or utilized in any form or by any means, electronic or mechanical, including photocopying, recording or by any information storage and retrieval system, without permission in writing from the publisher.

McKeown, USN, Captain Tom
Turn the Ship
A Justified Mutiny
1st edition

Library of Congress Control Number: 2021950736
ISBN 978-1-954604-05-6 (paperback)

Published by
AquaZebra
Cathedral City, California
www.aquazebra.com
Cover/interior design
Mark E. Anderson, AquaZebra

Printed in the United States of America.

Disclaimer

This book is nonfiction. It reflects the author's present recollections of experiences over time. Some names and characteristics have been changed, some events have been compressed, and some dialogue has been recreated.

Dedication

To my daughter Colleen McKeown and to
Carol Moore for her help in bringing this book to fruition.

Contents

Disclaimer...v
Dedication vii
Prologue.. xi
Introduction..xiii

Chapter 1	Sabotage..1
Chapter 2	The Mystery Unfolds 7
Chapter 3	Finally Underway.. 11
Chapter 4	The Island of Vieques..................................... 13
Chapter 5	Panama Canal to Hawaii 17
Chapter 6	En Route to San Diego 19
Chapter 7	Underway for Honolulu 23
Chapter 8	From the Philippines to Vietnam 27
Chapter 9	Message from Pentagon...................................35
Chapter 10	Joint Armed Forces Landing Exercise 43
Chapter 11	My Career on the Line 45

Prologue

A board of senior naval officers carefully screens and qualifies other officers for possible command at sea. This story is based on fact, in which an officer is selected for command of a U.S. Navy destroyer, someone who possesses introverted and narcissistic tendencies. The consequences are dire for the ship's crew, especially for a young officer, second in command, who willingly jeopardizes his career by rebelliously confronting the captain to avoid a perilous incident at sea.

After three months off the coast of South Vietnam, while the destroyer is providing shore bombardment support to U.S. Army and Marines fighting inland, the young executive officer receives surprising orders to take command of his own ship, an LST (Landing Ship Tank) homeported in Norfolk, Virginia. This disrupts the captain's evaluation of his mutinous executive officer. The remaining story is both humorous and intriguing as the new, young captain enjoys his command. However, he will wrestle with his conscience as to how and why he ever could have been given this command after his past mutinous actions on the destroyer.

Introduction

In January 1965, I received orders as Executive Officer (XO) of a destroyer. This is a major milestone for a surface line officer. The USS Robert Swanson DD 892 was homeported in Newport, Rhode Island. However, soon after reporting aboard, the chief of naval operations (CNO) ordered the ship to prepare for deployment off the coast of Vietnam.

It was wonderful being home with my family for two years while attending the Navy Post Graduate School in Monterey, California. Sailing off so quickly to Vietnam was not expected nor appreciated by me or my wife, Mary. After driving across country, we moved into Navy housing. We finished settling down and put our three boys and one girl in a new school.

Preparing the ship for combat included screening personnel, loading ammunition, conducting maintenance, and training. A change of command happened about two weeks before I reported for duty; a new commanding officer was in place. When I reported for duty, the new commanding officer was not onboard. He had decided to take off and spend time at home before we deployed. So as XO, I was in charge and did my best to acquaint myself with the officers and crew. I maintained a steady state of accomplishment as the date to deploy rapidly approached. Of course, I often phoned the captain and kept him informed. I was anxious to meet him, but without an invite, I remained steadfast at my job and went home to my family in the evening. I loved Mary so very much, and I adored my three little boys: Tommy, eight; Jim, six; Shawn, five; and my little girl, Colleen, two. The kids were involved in baseball and soccer, and of course the boys were always finding some way to get into trouble.

Chapter 1

Sabotage

After parking my car one morning and walking toward the ship, I was shocked to see a Shore Patrol car in front of the gangway, which was wrapped in yellow tape blocking the entrance. Approaching the gangway, the officer of the deck (OD) shouted to me, "XO, XO, Lieutenant Commander McGregor, come aboard! Hurry! We've been sabotaged!"

I showed my ID to the patrolman and he moved part of the yellow tape. I quickly ran up the gangway saluting the flag.

"What the hell is going on, O'Donnell?" I asked. "Why didn't you call me? Did you call the captain?"

"It happened five minutes ago, sir. All the wires controlling the turret for gun mount 52 have been cut."

"Holy shit!" I screamed. "I'll call the captain." Ensign Jim O'Donnell was a very competent, good-looking young officer about twenty-one years old, a recent graduate of Ohio State University. I trusted his judgement.

As Executive Officer (XO), I was second in command and traditionally expected to advise, defend and respond to the captain's commands and whims. I quickly phoned the captain and gave him a rundown as to what we knew so far. He told me he would be there in a while. "You take care of it," he said.

"Yes, sir." It's his ship. If anything goes wrong the ball drops in his lap. I was surprised at his lack of urgency in this situation.

As soon as the Officer of the Deck was told about mount 52, he was smart enough to call the shore patrol. They immediately blocked the gangway so no one could come aboard or leave, assuming it might be an inside job. I called the Naval Crime Investigative Service (NCIS). They arrived within an hour and were pleased we'd locked down the destroyer.

NCIS Senior Agent Bryson Allen introduced himself and asked me to take him to the gun mount. Allen was about six feet, five inches tall; a big, strong, stocky, black agent. He was all business. He carried a 45-caliber pistol in a leather case over his left shoulder, and he wore a gray, short-sleeve shirt with his NCIS badge in clear view.

As we were about to go to the gun mount, a gray Chevrolet drove down the pier and was stopped by a shore patrol officer. Jim O'Donnell told me, "XO, that's the captain with his wife driving the car."

I shouted to the shore patrol and the NCIS agents, "He's the captain! Please let him through."

The captain was a short, stubby-looking guy, about five foot five and maybe in his late forties. His face was quite wrinkled; probably a smoker for a long time. His wife pulled away without a kiss—like she was glad to get rid of him—and the shore patrol allowed him

Turn the Ship

through the yellow tape.

Navy Commander (CDR) Albert Coogler, our captain, had come up through the ranks. His last duty assignment was as executive officer aboard a destroyer tender in San Diego. He'd never had command of a ship. "I guess you're my new Executive Officer," he announced. "Tom McGregor, is it?" he asked with a grin.

"Yes, sir," I replied with a salute. He halfheartedly returned the salute, his hand barely reaching his hat.

"What the hell's going on?" he demanded.

"All the wires in the turret of Mount 52 have been cut. Chief Gunner's Mate Bob Long discovered this during his morning inspection. He then ran to the quarterdeck and informed Ensign O'Donnell, who was Officer of the Deck. O'Donnell called the shore patrol and notified the command duty officer, Lieutenant Sullivan. When the shore patrol arrived, they immediately blocked the gangway. When I arrived, I called you and then the NCIS. Let me introduce NCIS Agent Bryson Allen." Allen offered his hand to the captain.

"Obviously you're not a military man, or don't you know how to salute?" Allen looked down from his six-foot-five towering view at our plump, five-foot-two captain.

In a deep baritone voice, Allen explained, "I'm not an active military man, captain, and I must advise you that you will be interviewed before you can leave this ship. You are in lock-down mode like the rest of your crew." Allen continued, "I need to know the names of your people not on board. This will require the chief yeoman to identify those on authorized leave or liberty, and after that we'll have a 'sight' muster of all those onboard. Those who were off the ship during the incident will not be permitted to board the ship until my agents complete their jobs."

The captain was visibly shaken. Looking up, he nervously asked, "What kind of jobs are you talking about?"

Allen looked down at him and said, "The secretary of the navy has been informed of this incident. Captain, you need to be prepared to answer several questions, and you need to prepare an Operation Action Report."

The captain looked up at me and said, "You better get at it, XO."

Allen interjected, "The both of you, come with me to the gun mount and then to the wardroom for an interview."

Gun mounts on a destroyer

The captain and I followed Allen to the control turret of mount 52. Allen climbed the ladder and squeezed into the hatch. *Wow!* I thought, *He barely made it.*

Captain Coogler was next, and I had to shimmy his ass up. I looked in the hatch and saw no room for me, but I could see and hear the others. Allen pointed out the hanging wires. He explained that the cut wires were primarily around the control mechanism that moved the gun automatically up or down or port and starboard. "The gun is useless," he said. "You need to get this repaired in a hurry if you're deploying in a few weeks."

I went down the ladder and waited for them to climb out of the hatch. Captain Coogler's head came out first, like a turtle popping out of its shell. He wasn't supposed to come out that way, headfirst. He was supposed to put his arms out first, but I figured Allen pushed him up. I grabbed him and pulled him and his fat belly out the hatch. We both fell to the deck, but no harm was done. Allen came out the correct way and the three of us headed off for the wardroom.

I asked the wardroom steward to please make us some coffee and juice (affectionately called "bug juice," usually made with pigment powder from various fruits).

Allen started: "We don't know whether the wires were cut as an inside job, or by a stranger who came aboard." He took a sip of his bug juice and continued. "This is how NCIS will proceed. Two additional agents will soon join me to carry out the interviews. I must quickly interview the sailor who first discovered the disabled gun mount, and establish when he or anyone else last witnessed the gun mount working properly. We need to identify, as closely as possible, the time this incident occurred. With this knowledge, we will be able to dismiss most of the crew." Allen looked at me. "When have you been aboard recently?"

"I've been aboard every day over the past two weeks, from 0800 till 1800," I answered while looking at Captain Coogler.

Allen continued, "And you, captain, how often have you been aboard?"

Dropping his head sheepishly, Coogler replied, "Oh! Maybe two or three weeks ago."

"Okay," Allen answered with a frown. "You are dismissed, captain, and you may go home. I'll work with your XO, and I'm sure he'll keep you apprised."

Chapter 2

The Mystery Unfolds

"Lieutenant Commander McGregor, bring me the Chief Gunner's Mate who first discovered the problem."

"That will be Chief Bob Long," I said.

Long entered the wardroom. A real old timer with at least thirty years' service. Not a guy to mess around with. He wore a regulation gray beard and a handlebar mustache. His thinning hair was cut short. A slim old sailor with a mean attitude. "Yes sir," he said, "what can I do for you?"

"Have a seat, chief. This is NCIS Agent Bryson Allen who's in charge of investigating the Mount 52 incident."

Agent Allen stood up and reached out for the chief's hand. Chief Long seemed annoyed or nervous as he completed the handshake. "Chief, I need to know what time of day you entered Mount 52 and discovered the wires were cut?"

The elderly gunner said, "It was about 0930, during my morning rounds."

"When did you last see the wires intact?"

"About 1600 last evening. You don't think I cut the damn things, do you?"

"Chief," I said, "hold on. From what I hear, you're the oldest and most respected sailor on this ship. We need your help. I'd like to catch the bastard who did this.

"Mount 52 is one of my best guns," he responded, "and I take very good care of it and all the ship's guns."

Agent Allen stared at the mean-tempered chief for about ten seconds. "I don't have any other questions for you at this time, chief, but I need you to stay onboard."

Chief Long left the wardroom, grumbling to himself.

Allen turned and addressed me: "XO, I need all hands on deck for an eyeball-to-eyeball muster. I will then go to each division, explain to them what's happening, and get a list of those not present. This absentee list should match the total authorized absentee list of your yeoman."

Six divisions were on deck, and Allen was finishing the first division, about twenty-five sailors, responsible for the forward part of the deck, including the gun mounts. The division officer reported all present, except for Seaman Farley.

But then one of the sailors in the back row shouted, "Sir! I saw Farley in the forward head right before muster. He was talking with Chief Long. I'm sure he's on board."

Allen turned to him and replied, "Okay, sailor, go get him and bring him up here."

Seaman Apprentice Farley sheepishly joined the rear ranks. The division officer said, "Come up here, Farley. Why the hell didn't you report on deck for the eyeball-to-eyeball? Didn't they teach you that at boot camp?"

"Yes, sir! I mean, no, sir!" Farley was shaking, especially with the big NCIS agent staring him down.

"Come with me son. Division officer, I'm taking him to the mess hall."

Farley followed him through the passageways, down the ladder to the mess hall. They sat down at the table on seats facing each other. Allen could hardly fit. Farley began to tear up.

"Farley, what's your first name?"

"Alfred," he said.

"Do you work for Chief Long?"

"Yes," he mumbled, lips quivering.

"Did you ever go into Mount 52?"

"Yes, sir, and I liked to control the turret and swing the gun around, but Chief Long would yell at me. The last time, he told me never go in there again or he was going to throw me over the side."

"But did you?"

"Yes, sir." Farley began crying hysterically. "I went into the turret and moved the gun way off the center, and I couldn't get it to come back. I was so scared he would know it was me that I cut lots of wires to make him think the gun was broken. I don't know what he's going to do to me or what's going to happen to me."

"Alfred, I'm going to get you transferred off the ship to the naval station. You will face some disciplinary charges, but at your age they may not be too severe."

Agent Allen sent for me and explained the whole story. "LCDR McGregor, write up the charges on Farley and send him to the naval station. Let me see the charges. Do not use the word 'sabotage.' You can dismiss the men from muster, but do not tell anyone about Farley except your captain and his division officer. McGregor, I'm not coddling him, but I have some suspicions, and I need to go into that gun mount again. You take him to the naval station, executive officer."

Suddenly, Chief Robert Long came bursting into the mess deck. "There he is, that little son of a bitch. Let me get my hands on him!"

Allen stood up, blocked Farley, and told the chief, "Get the hell out of here. XO, keep your eye on his skinny ass. He's a menace." The chief slammed his hat on the deck, turned around, and sheepishly looked at Allen. Then he bent over and picked up the hat and left. Obviously, no one was going to mess with our NCIS deckhand.

Chapter 3

Finally Underway

I sent a well-done message to NCIS, and I commended the outstanding performance of Agent Bryson Allen. Not only was he thorough, but he was kind-hearted deep inside. I made a new friend, and we shared a few beers together at the Officer's Club.

It took us a very busy two weeks to get the gun mount repaired, load the ammunition, overhaul electronics and the engine room components, and transfer and welcome new hands. Farley was transferred to the Naval Base for disciplinary action and mental attention. To my chagrin, as we prepared to get ready for our journey, the captain seldom came aboard except to bring his clothes and other paraphernalia.

Trying desperately to figure him out, I soon lost respect for him when he kissed his dog goodbye and ignored his wife completely the day we were pulling out. No question; he was different. But he was the captain, and I kept trying to gain his confidence. It was difficult for me to say goodbye again to my wife and the kids. Mary was expecting our fourth baby, and we knew this was going to be a long deployment.

On the bridge, Captain Coogler loved to sit in his captain's chair and sleep, while I controlled the ship. We had a long journey from Newport, Rhode Island, through the Panama Canal, to Hawaii, the Philippines and then the coast of Vietnam. All the officers found it difficult to converse with Coogler. An introvert, I thought, which made my job and all our jobs more difficult.

Chapter 4

The Island of Vieques

We received a priority message from Commander in Chief of Destroyers Atlantic to proceed to the island of Vieques for shore bombardment practice. This was welcome news to me since we could check the status of Mount 52, as well as all our major guns.

The island of Vieques is located about seven miles south of Puerto Rico. It was set up as a firing range for U.S. Naval ships in the late 1960s. A few inhabitants were located on the far side of the island, safely away the ship's bombardment range. Still, it bothered me that human life may be in danger.

Before manning the guns, I asked the captain if a few of us officers could lower a lifeboat and go ashore onto this magical, pristine island to gather some conch shells and other items of nature on the beach. It would only take about an hour or two. He nodded his head. So Lieutenant Ambers, Chief Long and I lowered the boat and off we went.

As we drove toward the beach, the ocean was so clear you could see the bottom twenty feet down, but you could also see gun shells that had fallen well short of their targets. When we reached the shore, the waves were less than a foot high. It was a paradise of coconut palms and flowering cactus. A lush, unmatched natural beauty, untouched by human hands for several years. Conch shells and unspoiled sand greeted us.

The beach at the island of Vieques

After anchoring the boat we wasted no time gathering the conch shells and even some lobsters hidden in rock beds. Just as we were about finished loading the boat, we heard a loud *boom!* We took cover as gun shells flew over our heads. "That's our ship firing at us!" Chief Long yelled.

"What, is this guy nuts altogether?" I shouted. "Give me that ship/shore phone and grab an ore and start waving it!"

Luckily, I got hold of the fire control director and told him to ceasefire. "We're still on the beach!" I screamed.

He replied, "The captain ordered me to commence firing. He said you guys were on your way back."

"We are now," I said. "Oh, God, give me patience with this nut case."

Chapter 5

Panama Canal to Hawaii

The captain was in his glory as we were guided through the Panama Canal. The canal was one of two locations where the captain of any U.S. Navy vessel was not responsible for navigating his or her ship. Besides the Panama Canal, the other occasion is entering and leaving drydock at a shipyard. The captain ran all about the main deck, taking pictures and enjoying himself, but not speaking with anyone.

The Panama Canal is a man-made, fifty-one-mile waterway in Panama that connects the Atlantic and the Pacific Oceans. The canal cuts across the Isthmus of Panama and is a conduit for maritime trade. It greatly reduces the time for ships to travel between the Atlantic and Pacific, enabling them to avoid the lengthy, hazardous Cape Horn route around the southernmost tip of South America. Canal locks are at each end to lift ships up to Gatun Lake, eighty-five feet above sea level, and then to lower the ships at the other end. The original locks were 110 feet wide. We had a beam of about eighty-five feet. The Atlantic entrance is through Limon Bay, a large deep-water harbor with a port named Cristobal. We had to go dockside and wait for the pilot and our turn to enter. At Cristobal we were able to take on fuel, but our chief engineer was hesitant to top off because of the commercial quality of the fuel used by large merchant ships, not Navy destroyers.

Panama Canal

The entrance of the canal runs about five and a half miles. Then the Culebra Cut slices seven and a half miles through the mountain ridge. It crosses the Continental Divide of the Americas and passes underneath the Continental Bridge.

Finally, we were underway with the pilot giving the rudder and engine orders. The captain was delighted. I stayed on the bridge. After a long and exciting trip of up and down the locks, we reached the Miraflores Lake, one and a half miles long and fifty-four feet above sea level. From the Miraflores Lake, we descended into Balboa Harbor.

Chapter 6

En Route to San Diego

When we left the canal, Captain Coogler returned to his chair on the bridge, again not speaking to anyone—except to yell at me, "Take over, XO!"

I set a course to head north along Baja and the California coast. I convinced the captain that we needed fresh provisions and that we should pull into San Diego before heading 2,500 miles to Pearl Harbor. A cold beer for some of us wouldn't hurt either.

After two days at sea—remaining at least ten miles from the coast so we could pump bilges and offload the sludge accumulated from the commercial fuel taken on at Cristobal—we were finally approaching the main San Diego Channel Buoy. Unfortunately, it was at night. I suggested we wait for daylight, but the captain insisted we go into the harbor. It was about 0400 so I knew daylight would soon be here.

As we passed Point Loma, the red and green channel markers became difficult to distinguish from the flickering lights on the shore, mostly cars and billboards on the city highways. I slowed the ship and relied on radar. The chief quartermaster contacted harbor control, and we were assigned a berth portside on the 28th Street pier. I made a sharp turn to the right, passing North Island, and steadied on a course for 28th, cautiously avoiding the ferry crossing going to and from San Diego and Coronado Island. As we approached the pier, I ordered a sharp left turn to moor portside.

"Oh, my God!" the Helmsman yelled. "XO, we've lost control of the rudder!"

I shouted, "After Steering, take control!" and then I ordered back down full. After Steering is located below deck near the rudder itself; two sailors stand two-hour watches, ready to move the rudder manually.

I looked up and directly ahead of us was a Soviet freighter, flying its red flag with the crossed gold hammer and sickle. Most likely, it was permitted to dock for emergency repairs. Our ship was vibrating furiously as it was straining to back down. Our bow was preparing to slice into the Soviet hull. Russian sailors were running on their deck with tires and padded fenders to lessen the impact of our ship's bow stem. I closed my eyes; the captain put his head between his knees, screaming like a banshee, waiting for the crash.

Soviet freighter

Turn the Ship

I slowly opened my eyes and began feeling the ship miraculously moving backwards, still vibrating. The ship was now picking up speed going backwards. I had to kick it ahead full again so it wouldn't plow stern-first outside the channel into the mud. At last, our ship stopped its torturous maneuvers, and I was able to turn it and moor portside to the 28th Street pier. I could hear the Russian sailors clapping and our sailors whistling as we came to rest alongside the dock.

The captain sat back in his chair, wiping his brow. After the panic, I began to recall why we had stopped here in the first place. I was concerned about taking on fuel and fresh provisions and—mostly—getting a drink!

Everything was going smoothly. Hoses were connected to the pier drawing in fuel. Trucks were lined up, and our men were carrying fresh food and beverages aboard. So I grabbed Lieutenant Joe Ambers, and we took off for the San Diego Officers Club. As we approached the bar, I noticed two men in strange uniforms, grayish with black shoulder boards, sitting side-by-side, also at the bar. I looked at the bartender and he raised his eyebrows. Joe and I ordered a couple of beers.

"Good afternoon, gentlemen," I said.

They both turned toward us and replied "Мы не говорим по-английски." Obviously, they were from the Soviet freighter—the ship in which we nearly sliced a pie-shaped hole. We gestured, showing the right hand slicing into the palm of the left. We laughed together, finished our drinks and returned to our ship.

Chapter 7

Underway for Honolulu

We all looked forward to Hawaii, knowing it was our last U.S. liberty port before reaching Vietnam. It took at least five full days at sea, nearly 2,500 miles, before we saw the tips of the eastern islands. We also saw wayward seagulls hoping to find food from this welcome vessel.

Soon we approached the Pearl Harbor channel buoy. From there, we set course and entered the channel, passing the Honolulu International Airport on our right. Parallel to the commercial airport was the Hickam Air Force runway, which, on occasion, was used for either commercial or military landings.

Prior to entering the harbor, I ordered all hands on deck and to man the rails in their white uniforms. When we passed the Arizona Memorial, we all came to attention and saluted. A tearful moment was experienced by most of us.

Pearl Harbor control assigned us a berth at the naval station. Soon afterward, the ship lines were tossed over to waiting sailors who tied them to the bollards, and we were secured to the pier.

Arizona Memorial, Pearl Harbor

The next morning, we were shocked to see a local police patrol car and a black NCIS Hawaii van pull up to our gangway. The officer of the deck called me, and I was surprised to see the same burly agent who interrogated us in Newport, Rhode Island—Bryson Allen. At the top of the gangway he saluted the flag flying at the stern and then turned to me, "You are Lieutenant Commander McGregor, we meet again. Is the captain aboard?"

"Yes, sir. I'll take you to him."

"I have new evidence regarding the sabotage incident in Newport, Rhode Island."

After knocking on the captain's door, he opened it and recognized the imposing NCIS agent. "It's you again," he said. "Are you on vacation?"

"No, sir. NCIS has an office here in Hawaii, and once again I have important business to attend to on your ship," Allen replied.

Turn the Ship

"Can you come to the wardroom with me?"

"Of course."

Once we were in the wardroom, the three of us sat down and ordered coffee. The NCIS agent began speaking. "Gentlemen, new evidence has cleared the young man who was forced to confess to cutting the wires of Gun Mount 52. It was not the work of young Seaman Apprentice Farley. We have solid evidence it was Chief Petty Officer Long. He was paid by an international spy team."

Chief Petty Officer Long was called to the wardroom and told to put his hands behind his back. Agent Allen handcuffed him and escorted him off the ship to await his fate.

After this incident, I passed the word that all hands could start liberty which would expire at 0600 the next morning. The captain once again chose not to leave the ship.

All the time we were in Hawaii, our reclusive captain had never left the ship.

On our way to Hawaii, I first recognized his vindictiveness. About a day out from reaching the islands, a young Ensign reported to his watch one hour late. The captain put the young man in hack (restricting him to the ship) and refused to let him ashore. I pleaded with him to give the ensign a break. He refused. The young lad had never been to Hawaii. However, before leaving Hawaii, I smuggled a paper cup Mai Tai to him. The rest of us were underway with Aloha hangovers.

Chapter 8

From the Philippines to Vietnam

During our ten days at sea traveling to the Philippines, I often met with the chiefs. From them, I learned how the captain frequently disparaged me. It's the chiefs who run the Navy and this news was depressing. Hell, this was my career. What more could I do for this guy? He stayed in his room or slept on the bridge. And he didn't like to talk.

I brought the ship into U.S. Naval Base Subic Bay, located in Zambales, Philippines, for provisions.

I requested permission from the grinch to go ashore. I asked if he would like to join me. He shook his head. I grabbed my golf clubs and called a cab. Two chiefs, my operations officer and I headed for the nearest golf course. On the sixth hole, I was about to putt when a sailor came racing down the fairway in a golf cart screaming, "XO! XO! The captain wants to get underway immediately. He sent me to get you because there is a typhoon heading our way."

U.S. Naval Base Subic Bay, Philippines

"Okay, okay, let me finish my putt!"

When I returned to the ship, the wind and rain were already howling. The captain was a nervous wreck. "There's a typhoon heading our way. We have orders to circle the Paracel Islands, and I want to get moving," the captain demanded.

"Why are we even getting underway with such terrible weather?" I said.

The captain repeated "We must get underway to circle the Paracel Islands." He insisted we get going. "There's a typhoon heading our way. We have orders to circle the Paracel Islands, and I want to get moving," the captain demanded.

I went to my stateroom and put on my rain gear and a life jacket. I then went up to the bridge and ordered all there to put on their life

Turn the Ship

jackets. The captain looked at me and was frowning. Nevertheless, he also put on a life jacket.

After finally rounding up the crew, I repeated my orders for all hands to prepare for rough weather and to put on their life jackets. We took in the lines and proceeded out the channel—much to my chagrin.

One day later, Chief Quartermaster O'Malley, who navigates the ship advised me that we were headed directly into a 200-mph typhoon. "XO, we must turn back and head for shelter around the Philippine Islands!" he exclaimed.

"Have you told the captain?"

"He won't listen to me."

I decided to talk to him.

"Damn it, we're going to the Paracel Islands, XO. Don't be such a wimp!" he shouted.

This guy was nuts. We had to turn the ship. I decided to bring four officers and Chief Quartermaster O'Malley to the bridge and confront the captain. He was horrified to see this crowd on the bridge.

"What the hell's going on, XO?"

In a loud and determined voice, I instructed all the officers to listen to me and the Chief Quartermaster to log all that was said. The captain was shocked.

I said, "Captain, this ship and our crew are heading for total disaster. We must turn 180 degrees and head for shelter behind Luzon at full speed."

"XO, this is mutiny. Your career is ruined. I'm giving you an UNSAT fitness report. Your career is over."

The 200-mph typhoon

"Quartermaster, log it in. Captain, turn the ship!" I pointed a finger and insisted. "Turn the ship!" He came out of his chair walked around staring at each of us.

Grumbling and swearing. "All right, all right," he finally acquiesced, "turn the damn ship."

Turn the Ship

Two weeks later we circled the Paracel Islands, south of Hainan, China, and eventually took our station off the Vietnam coast. We were always dressed in battle gear: helmets, shirts buttoned to the neck, and life jackets. We were there in the summer months, hotter than hell. Our ship was assigned a call for fire mission supporting our troops in country. The Marines or Army would send us the coordinates. Fire control would train the guns. The captain or I would give the order to fire.

We also had a ten-mile-square sector at sea starting ten miles along the shoreline and then out ten. Day after day we slowly patrolled our sector, responding to the troops and looking for sector intruders. About every three days, a Navy tanker or aircraft carrier would come our way. We would refuel, get some fresh provisions and send and receive mail. We let our beards grow. Our orders were to shoot anything that entered our sector.

Then it happened!!

Through bloodshot eyes one of our lookouts spotted a sampan being dragged into the water. Adjusting my binoculars, I could see an old man tugging on his boat trying to get it waterborne. He probably wanted to catch some fish. I reported the sighting to the captain.

"Notify all guns to prepare to fire," said the captain

"Captain, he's just a little old man who wants to go fishing."

"Our orders are to shoot anything," he said with a smirk. "Now shoot his ass, XO. That's an order."

I grabbed my intercom and told Fire Control to line Mount 52 battery on the sampan and fire at my command.

The captain screamed, "I want all five-inch guns on that target! Wipe him out!"

Chinese River Fisherman

That's three turrets, each with twin five-inch caliber, circular-barreled guns capable of sinking a battleship, or supporting our troops with explosive projectiles over six miles inland. Now we were going to use all that fire power to shoot an old man in a sampan?

"XO, this is Maguire in Fire Control. All guns are on target, ready to fire, sir."

By this time the sampan was at the water's edge and the old man was jumping in.

"Maguire, from XO, Fire!"

The ship rolled port to starboard with a deafening blast. I looked toward the shore. All the projectiles went over his head and hit the rice paddies, eliminating tons of stones and almost all the hill. I'm sure the little old guy didn't know what the hell had happened.

"Did you get him?" shouted scrooge as he rubbed his hands together.

"No, sir," I responded angrily.

"Keep firing, we have our orders. I want to kill him."

The crew on the bridge looked at me with raised eyebrows.

"Fire Control, this is the XO, standby to fire."

"Aye, aye, sir," Maguire snapped back.

The three turrets began squeaking as they adjusted their twin guns. The sampan was now in the water.

"XO, this is Fire Control, ready to fire."

"Very well . . . fire."

This time, all the rounds came up short, crashing into the water close to the sampan. Huge waves caused by the gun's projectiles picked up the sampan tossing the old man and his boat into hills. I saw him crawl out of his splintered boat and run into the forest.

"Did we get him, XO?"

"Yes, sir, got him good."

Thank God the poor bastard was alive. No one on the ship contradicted me. The captain immediately sent a message off to the admiral explaining how he interdicted and destroyed a boat carrying weapons for the Viet Cong.

We resumed our patrols. The monsoon season began, and the rain felt good as we suffered under the heat. It was about three months now since I was accused of mutiny and demanded that the captain "turn the ship." I began wondering what I was going to do when I left the Navy after the Pentagon received that fanatical fitness report. A career scuttled trying to avert disaster.

Chapter 9

Message from Pentagon

One hot rainy night, I turned the bridge watch over to our operations officer, went below and flopped on my bunk, wet clothes and all. I fell asleep at once. Who needs to shave and shower?

Around three in the morning, a loud knock at my stateroom door scared me awake.

"Okay, okay. What the hell do you want?" I growled.

"Sir, its Radioman D'Angelo. I have a message from Washington for you."

"For me?"

"Yes sir."

"All right, let me see it."

I read the message three times. "D'Angelo, if this is some kind of a joke, heads are going to roll."

> Lieutenant Commander McGregor,
> Congratulations. You are to take command of the USS

Monmouth County LST 1289 on or about April 26, 1968, at Norfolk, Virginia.

Wow! What a mistake this was. They hadn't looked at my records.

"D'Angelo, has the captain seen this yet?"

"Yes, sir, and he's pretty angry. He wants to see you immediately."

I pulled myself together and followed D'Angelo to the bridge. There he was, sitting in his chair as usual.

"McGregor have you seen this message?"

"Yes, sir." I answered.

"Well, take a look at my reply."

> Lieutenant Commander McGregor is the most insubordinate officer I have known. He has confronted me in a near mutiny manner. I have given him an unsatisfactory fitness report. He does not deserve command.

I responded, "Thank you very much, sir."

He laughed. The bridge crew turned away from him.

I went back to my stateroom. The thoughts were circling in my head: *It's all over. I'm done.* Unable to sleep, I returned to the bridge at sunrise.

It was a beautiful sunrise. A smooth, nearly flat sea with the rising rays from the sun streaking across the sparkling water like a giant fan. I loved going to sea. I would miss it.

Suddenly, D'Angelo taps me on the shoulder. "XO, you must see this latest message from the Pentagon."

Turn the Ship

"What is it, my walking papers?"

D'Angelo smiled.

> Your message regarding the performance of Lieutenant Commander McGregor was reviewed. Carry out your orders. Lieutenant Commander McGregor is to be transferred in time to take command of USS Monmouth County LST 1289 on or about April 26, 1968, at Norfolk, Virginia.

I decided not to discuss this with the captain. There must be a mistake somewhere. All I thought of was going home to my wife and kids. I wonder if they were sending an officer to relieve me.

We received another call for fire from the marines. Fire control was ready. Mounts 52 and 53 loaded up, and I gave the order to commence fire. A few minutes later we heard from the Marines: "On target! Bingo!"

We all cheered. Another hot, muggy day doing our job.

Later that day, the captain sent for me. He was in his cabin. "McGregor have you seen all these messages from Washington?"

"Yes, sir," I acknowledged.

"Well, I just got another one. Lieutenant Commander Abrams is going to relieve you. He's on his way. He will be airlifted to the Enterprise. I'm sending you to the Enterprise by helo soon after he arrives. I'll be happy to see you go."

I felt like saying many things, but I just gave him the old, "Yes, sir."

Time seemed to drag by, and my excitement was peaking. Getting close to going home, I was very careful and hoped that nothing eventful would happen. Our biggest threat was the North Vietnamese, Russian-made MiGs. When they showed up, USS Enterprise launched fighters and the MiGs scattered. Occasionally, we fired our 20mm guns when ordered. They never came close. However, we did see several of our fighters take them out. It was another reason to cheer.

Finally the day came. We were on our way to rendezvous with the Enterprise.

I had one of our young officers bring the ship alongside the carrier and take a station about eighty feet between the ships. He did a fine job, and I tried not to breathe down his neck. We began firing our lines over to the carrier. Their sailors would pick them up and pull them over. It took about ten men for each line. A signal light on the Enterprise began to blink. Our signalmen responded with their lights.

I delivered the message to the captain, "XO Lieutenant Commander Abrams is aboard, and they're going to highline him over."

"Is Abrams aboard the Enterprise?" the captain shouted.

"Yes, sir. He's going to be highlined." I replied.

Being highlined from ship to ship is a thrilling experience. You put a lifejacket on and sit down in a metal cage that's open in the front. The cage is attached to a manila line with a pulley, and the crews on both ships give and take the line until the cage gets across. The ships steaming along so close create a raging sea. Looking down into sea is not a good idea. They put Abrams in the cage, and I prayed for him. When he reached the midpoint, the two ships rolled in opposite directions. Up popped Abrams, high above the ships, rolling over and over.

"Oh, shit!" I blurted out. "There goes my relief."

Then Abrams was dropped in the water as the ships rolled toward each other. Then he was yanked up again. I could hear him scream; I screamed too. The whole scene reminded me of a toy that, when squeezed, a monkey would flip over.

A soaking wet Abrams finally made it across. He was escorted to the bridge. I gave him a bear hug.

"Nice ride," I said. "Welcome aboard. They sent your duffle bag over earlier. Go down and shower."

"I'd rather be shot than ride that thing again. Thanks."

We began taking in the lines. My young officer was anxious to kick it up and take off. I cautioned him not to turn away too sharply or our wake would collide with the carrier's bow. Now we were off and running with an impressive rooster tail proudly displayed at the stern.

"Where is Abrams?" the captain shouted.

"He was soaking wet, so I sent him to change. He'll be right back," I said.

"He is supposed to report to me, not you. Send for him. I want to see him immediately."

Not a good start for Abrams.

Bob Abrams and I shared experiences. I warned him of the idiosyncrasies of the skipper. I told Abrams how the captain disdained me and how he was going to depend upon him to run the ship. I introduced him to the officers and chiefs, went through the files and took him to the bridge. Fortunately, Bob had served on several destroyers and was familiar with ship handling. We informed the captain that Abrams was ready to take responsibility as XO.

"Very well," he said. "I'm sending for a copter, and you're on your way tomorrow, McGregor."

I packed, jamming most of my essentials into a duffle bag. I packed papers, desk items, and some reports in a box and gave Abrams some money to mail it to USS Monmouth County. I don't think I slept a wink that night. I was on the fantail at 0600, ready to go. I saluted the captain who returned a sheepish grin. As the helicopter approached, I noticed almost half the crew lining the rails to bid me farewell.

Chief Quartermaster O'Malley and D'Angelo joined me on the fantail. The helicopter hovered above and lowered a line. They attached my duffle bag and up it went. Next, they lowered an orange yoke for me. The chief put it over my head and showed me where to put my arms.

"XO, we're going to miss you. Have a safe trip. Congratulations again," said O'Malley.

Then, while being hoisted, the crew started applauding. My eyes filled up. This was one hell of a sendoff.

Four days later I arrived at Logan Airport, Boston. My wife, Mary, and our five kids were in the waiting area as I came down the ramp. They all ran to me. Mary and I enjoyed a long, overdue kiss. I gave all the kids kisses. Then I picked up my two little girls, and we headed for a taxi.

Turn the Ship

It was along ride to Newport, Rhode Island. Everyone had stories to tell, especially my three boys who had to tell Dad about Little League baseball and football. When we arrived at our quarters, signs were all over the lawn. "Welcome home, Tom." "Welcome home, Dad." Our neighbors came out to greet me. Tears flowed again.

I had thirty days' leave. Normal for duty transfers. We traveled to New Jersey to visit grandparents and relatives, and the kids had a great time. Unfortunately, we had to think about moving to Norfolk, finding a house and new schools, and sadly saying goodbye to our Newport friends. This was especially hard for the children.

My mother watched the kids as Mary and I traveled to Norfolk to find a place. We arranged for a realtor to meet us. After seeing several houses, we found the one we wanted in Virginia Beach. We had movers pick up our belongings and truck them to our new home. By mid-April we were pretty much settled in, and I decided to report to my new command to see what it was like.

My command. Wow!

As I walked down the pier to my new command, I could not believe the size of the ship. About thirty feet high and 470 feet long, with huge bow doors that could swing open, launching trucks or tanks onto the beach.

I could see some sailors at the rails looking down at me as I approached the gangway.

Landing Ship Tank (LST) with its bow doors open

"Do you think he's our new skipper?" one of them asked.

"He's too young," said another sailor.

I was thirty-six and looked twenty-six. That wasn't always an advantage. Captain Robbins, the departing skipper met me at the quarterdeck. "Welcome aboard, Tom," he said. "Let's get a cup of coffee."

I was so excited. It was like Christmas morning. Nevertheless, deep inside of me was this awful feeling. This must be a mistake.

After a week of indoctrination, we were ready for the Change of Command Ceremony. My parents and all our relatives came for the ceremony. I know my mom and dad were very proud. It's a traditional and very formal event where the responsibility of command is clearly passed from one officer to another. I was overwhelmed with pride and confidence. I had the right experience and was capable. I knew that, but the awful feeling of whether this euphoria would last was painfully digging at me. During the reception, I was a good actor.

Three months went by. I took the ship to sea and had a ball. I even sat in the captain's chair, but only occasionally. My XO and I became very close. He was about ten years older than I was. He had been promoted through the ranks, from enlisted seaman to chief warrant officer to ensign and now lieutenant. I later found out that when we were in port and I would leave the ship and liberty would normally begin, the XO would pass the word on the mic "There will *be no* liberty until I give the word." He gained the nickname of "Be No."

Chapter 10

Joint Armed Forces Landing Exercise

One day our ship got orders to join a major military exercise at Camp Lejeune. This exercise included all the military services. An elaborately decorated grandstand was erected for dignitaries and guests of dignitaries.

As usual, Navy SEALs marked spots on the beach for landing craft and for my Landing Ship Tank (LST). Our ship carried several trucks, amphibious vehicles and a platoon of marines.

The Air Force began the exercise by attacking targets in the hills behind the beach. The army followed up, emptying soldiers on the beach from small Navy landing craft.

The overall admiral in charge sent a signal to our ship to proceed to the beach. I ordered full speed ahead and to standby to drop the stern hook. As we approached the shoreline, we dropped the stern hook only to have it mired in a sandbar which caused the stern line to break. So off we went to the beach at full speed.

Grandstand with guests and dignitaries

Our flat bottom ship rolled up on the sand, which seemed like it was made of ball bearings, doing nothing to slow us down.

All the dignitaries in the grandstand quickly evacuated fearing for their lives. My chief engineer called me from the engine room, yelling that trees and bushes were passing by the portholes! He asked, "Where are we going Captain?"

We finally stopped inland, high and dry. The admiral in charge came up on the radio laughing and saying, "Nice job, Tom!"

Two days later at the high tide, three tugboats dragged us off the beach.

Chapter 11

My Career on the Line

During my command I often wondered how I was selected to be the captain of this ship, knowing I was possibly charged for mutiny by my previous, entertaining captain, and now I'd accidentally run my LST up the beach!

I decided to travel to Washington, DC, and visit the Pentagon so I could check my record and come clean with the Navy and my conscience. I purposely stole the last three months to relish the feeling of having my own ship. I went to the personnel department and asked for my records and many fitness reports. I sat down took a deep breath and began looking for my UNSAT fitness report from Scrooge.

After going through all my records over the past fifteen years, I still couldn't find my last terrible report. Now everything made sense. That report was missing. I was given a command by default. I must tell the admiral. I couldn't live with this lie anymore. I walked to the office of the Chief of Naval Personnel. I asked his secretary if I could see him for a few minutes.

"Come on in," he said in greeting. "What can I do for you?"

"Admiral," I began, "I went through all my records. I was told I was given an UNSAT fitness report, but I can't find it."

"What ship were you on?"

"The USS Robert Swanson DD 892, sir."

'What's your name?"

"Tom McGregor, sir."

"So, you're McGregor!" he exclaimed.

The Admiral reached into the bottom drawer of his desk. With two hands, he lifted a huge pile of letters and threw them on the desk.

"McGregor, these are letters from your crew, all telling me how you put your career on the line to save the ship and their lives from that typhoon."

He stood up, reached for my hand and shook it.

"The Navy threw that bad fitness report away," he said. "I am proud to meet you. Go enjoy your command. You deserve it. Good luck and smooth sailing, captain."

If you enjoyed *Turn the Ship*, be sure to pick up Captain Tom McKeown's first book *Londonderry Farewell* ... wherever great books are sold.

www.ingramcontent.com/pod-product-compliance
Lightning Source LLC
Chambersburg PA
CBHW031429040426
42444CB00006B/747